7/01

STATISTICAL THEORY

The Relationship of
Probability, Credibility and Error

To
R. F. W.

LANCELOT HOGBEN, F.R.S.

Statistical Theory

The Relationship of
PROBABILITY, CREDIBILITY
AND ERROR

AN EXAMINATION OF THE CONTEMPORARY
CRISIS IN STATISTICAL THEORY
FROM A BEHAVIOURIST VIEWPOINT

W · W · NORTON & COMPANY INC
Publishers   New York

PRINTED IN GREAT BRITAIN

CONTENTS

FOREWORD

THE USE of the word behaviourist in my sub-title calls for clari-
fication. It came into current usage chiefly through the writings
of J. B. Watson, in the first fine flush of enthusiasm following
the reception of Pavlov's work on conditioned reflexes. Watson
conveyed the impression that all forms of animal—including
human—behaviour are ultimately explicable in terms of neural
or humoral co-ordination of receptors and effectors. Neither
this proposition nor its denial is susceptible of proof. A comet
may destroy the earth before we have completed a research
programme of such magnitude as the operative adverb *ultimately*
suggests.

Many years ago, and in what now seems to me a very im-
mature volume of essays* written as a counter-irritant to the
mystique of Eddington's *Nature of the Physical World*, I suggested
a more restricted definition of the term and offered an alter-
native which seemingly did not commend itself to my contemporaries. By *behaviourist* in this context I mean what Ryle
means in the concluding section of his recent book *The Concept
of Mind*. What Ryle speaks of as the behaviourist viewpoint and
what I had previously preferred to call the *publicist* outlook has
no concern with structure and mechanism. Our common
approach to the problem of cognition is not at the ontological
level. The class of questions we regard as profitable topics
of enquiry in this context arise at the level of epistemology,
or, better still, what G. P. Meredith calls *epistemics*. If one
wishes to discuss with any hope of reaching agreement what
one means by knowledge—or preferably *knowing*—the two
main topics of the agenda for our *public* symposium will be
recognition and *communication*; and we shall discuss them as
different facets of the same problem, neither meaningful in
isolation. A simple illustration will suffice to make clear, both
at the emotive level of non-communicable private conviction
and at the public or communicable level of observable behaviour, the difference between *knowing*, conceived as a process,
and *knowledge*, conceived as a static repository.

* *The Nature of Living Matter* (1930).

7

We shall suppose that B is colour-blind to differences in the red and green regions of the visible spectrum, A being normal in the customary sense of the term in the relevant context. Faced with a crimson Camellia and a blade of grass B asserts that they differ only with respect to light and shade, shape and texture, possibly or probably also with respect to chemical composition, microscopic structure and other characteristics less accessible to immediate inspection. A admits everything in this assertion except the implications of the word *only*. He *knows* that red is red and green is green, just as surely as B *knows* that the distinction is frivolous. The deadlock is complete till they call in C who suggests that it is possible to submit the subject-matter of their disagreement to the outcome of an identity parade. At this point, we shall assume that A and B, each equally secure in his convictions, both agree to do so. The proposal C offers takes shape as follows. First, each disputant inspects without handling a set of objects of like shape arranged in a row. Some are green and some are red. A labels the red ones $R_1, R_2, R_3 \ldots$ etc., and the green $G_1, G_2, G_3 \ldots$ etc. B denies that they are distinguishable. The umpire C now photographs the set twice in the presence of A and B, using first an Ilford filter which cuts out all rays in the green region and then one which cuts out all rays at the red end of the spectrum. He invites both A and B to watch while he develops and fixes the film. The reader may complete the rest of the parable. All that remains is to point out the moral.

If we are content to discuss the contemporary status of the calculus of probability as an instrument of research in terms of what goes on *in the mind*, we have as little hope of arriving at agreement as had A and B before the advent of C. If we seek to find a highest common factor for the process of knowing, we must therefore find some means of submitting our differences to an identity parade. We shall then cease to chatter in the private idiom of conviction and begin to communicate in the public vocabulary of events. Needless to say, we cannot force either B or A to do this. We can merely invite them to do so. As Ryle would put it, the utmost we can do in the context of the topic of the pages which follow is to make *linguistic recommendations* which will help them to resolve their differences, if willing to do so. At the outset, we must therefore insist on some acceptable distinction between *knowing* and *believing*.

The behaviourist, as I here use the term, does not deny the convenience of classifying *processes* as mental or material. He recognises the distinction between personality and corpse; but he has not yet had the privilege of attending an identity parade in which human minds without bodies are by common recognition distinguishable from living human bodies without minds. Till then, he is content to discuss probability in the vocabulary of *events*, including audible or visibly recorded assertions of human beings as such. He can concede no profit from a protracted debate at the level of what goes on *in* the mind. The use of the italicised preposition in the idiom of his opponents is indeed inconsistent with any assertion translatable in terms of a distinction C can induce A and B to inspect; and this dichotomy is not one which concerns the mathematician alone. It is not even an issue on which prolonged preoccupation with intricate symbolic operations divorced from contact with living things necessarily equips a person to arbitrate with a compelling title to a respectful hearing. None the less, it is one which forces itself on our attention from whatever aspect we view the contemporary crisis in Statistical Theory.

Hitherto, the contestants have been trained mathematicians, many of whom explicitly, and many more of whom implicitly, reject the behaviourist approach outlined in the preceding paragraphs. The writer is a trained biologist. Like most other trained biologists, he has a stake in knowing how far, and for what reasons, statistical theory can help us to enlarge our understanding of living matter. Since mathematical statisticians have not been hesitant to advance ambitious claims about the help it can give them, it is not presumptuous for a biologist to examine the credentials of such claims in the language of behaviour, the only language which all biologists share. If statisticians wish to decline, as Carnap most explicitly does decline, to communicate with the biologist on this level of discourse, they are still free to do so on this side of the Iron Curtain; and, happily, there are publishers who will make their views accessible to the consumer. It is entirely defensible to formulate an axiomatic approach to

*

the theory of probability as an internally consistent set of propositions, if one is content to leave to those in closer contact with external reality the last word on the usefulness of the outcome.

What invests the present controversies concerning the foundations of theoretical statistics with a peculiar flavour of comedy is a hyper-sensitive anxiety to publicise the relevance of the outcome to the world of affairs on the part of those who repudiate most vigorously the relevance of the external world to the choice of their axioms; but it is by no means true to represent the current crisis in statistical theory as a contest between opposing camps in one of which all the contestants are Platonists, in the other all of them behaviourists as defined above. The truth is otherwise. In one sense, we are all behaviourists nowadays. At least, few persons outside a mental home would defend thoroughgoing Platonic idealism. What is equally true is that all of us carry over from the past some remnants of mental habits which antedate the naturalistic outlook of our times. Hence we may consistently assert the relevance of the dichotomy to the content of the contemporary controversy while expecting to encounter among the disputants no greater measure of consistent adherence to one or other viewpoint than experience of the inconsistencies of gifted human beings can justify. Even Carnap, whom I have cited as a protagonist of the axiomatic school, does so with a generous concession to his own day and age by distinguishing between probability 1 and probability 2, or in my own idiom private and public probability.

This book is a prosaic discussion about public probability. If the topic were private, the author would have communicated his convictions through the medium of verse composition. One concern of the writer in the course of revaluating his own views by tracing current conflicts of doctrine to their sources has been to remedy an increasingly widespread disposition of the younger generation of research workers to relinquish the obligation to examine the credentials of principles invoked in the day's work. Indeed, one may well ask whether a liberal education in theology is nowadays less conducive to acquiescence in authoritarian dogma than is rule of thumb instruction in statistical methods as now commonly given to students of biology and the social sciences. Such acquiescence is no doubt due in part to the formidable algebra invoked by the theoreticians of statistics during the inter-war decades. For that reason, I have attempted in an earlier book to exploit a novel technique of visual aids to make accessible to a wider audience than those to which professional mathematicians commonly address themselves, the formal algebra of sampling distributions frequently referred to in this book. I trust that *Chance and Choice* will still serve some useful purpose on that account; but common honesty compels me to repudiate its claim to have accomplished an overdue revaluation of statistical theory.

In carrying out the self-imposed task attempted therein, my own views about widely accepted teaching became more and more critical as examination of current procedures against a background of explicitly visualisable stochastic models disclosed factual assumptions too easily concealed behind an impressive façade of algebraic symbols; but my concluding remarks were tentative and will have left the reader with what I now regard as an unduly optimistic anticipation of legitimate compromise. Here my aim has been to take stock of the outcome and to make explicit a viewpoint much less conservative than that of *Chance and Choice*. My present views are in direct contrariety to many therein expressed in conformity with current tradition.

Thus this book is not a textbook, complete in itself as a course of study. On the other hand, avoidance of any exposition of relevant algebraic expressions would have been inconsistent with the intention stated. To be sure, the lay-out would have been more pleasing, if it had been advisable to allocate more equally space devoted to historical and methodological commentary on the one hand and to algebraic derivations on the other. The writer has resisted the temptation to sacrifice the dictates of clarity to the aesthetics of book-making, because it has seemed best to steer a middle course between encouraging the reader to take too much on trust and recapitulating demonstrations to which he or she has now ready access. Thus the reader who commendably declines to take the algebraic theory of the *t*-test on trust may refer to Kendall's *Advanced Theory* or, if unable to follow all the steps set out therein, may consult Chapters 15 and 16 of Vol. II of *Chance and Choice*.

Contrariwise, Chapters 6–9 of this book have a more textbook flavour than most of those which precede or follow. I have written them in this way because biologists and sociologists will find little if any reference to the Gaussian Theory of Error in statistical textbooks addressed to themselves. Accordingly, they will not readily retrace the ancestry of current procedures subsumed under the terms regression and multivariate analysis unless forearmed by an explicit exposition of the method of least squares in its first domain of application.

CHAPTER ONE

THE CONTEMPORARY CRISIS
OR THE UNCERTAINTIES OF
UNCERTAIN INFERENCE

IT IS NOT without reason that the professional philosopher and the plain man can now make common cause in a suspicious attitude towards statistics, a term which has at least five radically different meanings in common usage, and at least four in the context of *statistical theory* alone. We witness on every side a feverish concern of biologists, sociologists and civil servants to exploit the newest and most sophisticated statistical devices with little concern for their mathematical credentials or for the formal assumptions inherent therein. We are some of us all too tired of hearing from the pundits of popular science that natural knowledge has repudiated any aspirations to absolute truth and now recognises no universal logic other than the principles of statistics. The assertion is manifestly false unless we deprive all purely taxonomic enquiry of the title to rank as science. It is also misleading because statistics, as men of science use the term, may mean disciplines with little connexion other than reliance, for very different ostensible reasons, on the same algebraic tricks.

This state of affairs would be more alarming as indicative of the capitulation of the scientific spirit to the authoritarian temper of our time, if it were easy to assemble in one room three theoretical statisticians who agree about the fundamentals of their speciality at the most elementary level. After a generation of prodigious proliferation of statistical techniques whose derivation is a closed book to an ever-expanding company of avid consumers without access to any sufficiently simple exposition of their implications to the producer-mathematician, the challenge of J. Neyman, E. S. Pearson and Abraham Wald is provoking, in Nietzsche's phrase, a transvaluation of all values. Indeed, it is not too much to say that it threatens to undermine the entire superstructure of statistical estimation and test procedure erected by R. A. Fisher and his disciples on

13

the foundations laid by Karl Pearson, Edgworth and Udny Yule. An immediate and hopeful consequence of the fact that the disputants disagree about the factual credentials of even the mathematical theory of probability itself is that there is now a market for textbooks on probability as such, an overdue awareness of its belated intrusion in the domain of scientific research and a willingness to re-examine the preoccupations of the Founding Fathers when the topic had as yet no practical interest other than the gains and losses of a dissolute nobility at the gaming table.

Since unduly pretentious claims put forward for statistics as a discipline derive a spurious cogency from the protean implications of the word itself, let us here take a look at the several meanings it enjoys in current usage. First, we may speak of statistics in a sense which tallies most closely with its original connotation, i.e. figures pertaining to affairs of state. Such are *factual* statistics, i.e. any body of data collected with a view to reaching conclusions referable to recorded numbers or measurements. We sometimes use the term *vital* statistics in this sense, but for a more restricted class of data, e.g. births, deaths, marriages, sickness, accidents and other happenings common to individual human beings and more or less relevant to medicine, in contradistinction to information about trade, employment, education and other topics allocated to the social sciences. In a more restricted sense, we also use expressions such as vital statistics or economic statistics for the exposition of *summarising* procedures (life expectation, age standardisation, gross or net reproduction rates, cohort analysis, cost of living or price indices) especially relevant to the analysis of data so described. By analysis in this context, we then mean sifting by recourse to common sense and simple arithmetical procedures what facts are more or less relevant to conclusions we seek to draw, and what circumstances might distort the true picture of the situation. Anscombe (*Mind*, 1951) refers to analysis of this sort as statistics in the sense in which "some continental demographers" use the term.

If we emphatically repudiate the unprovoked scorn in the remark last cited, we must agree with Anscombe in one particular. When we speak of analysis in the context of demography,

we do not mean what we now commonly call *theoretical statistics*
What we do subsume under the latter presupposes that our
analysis invokes the calculus of probabilities. When we speak
of a calculus of probabilities we also presuppose a single formal
system of algebra; but a little reflection upon the history of the
subject suffices to remind us that: (*a*) there has been much
disagreement about the relevance of such a calculus to every-
day life; (*b*) scientific workers invoke it in domains of discourse
which have no very obvious connexion. When I say this I wish
to make it clear that I do not exclude the possibility that we
may be able to clarify a connexion if such exists, but only if we
can reach some agreement about the relevance of the common
calculus to the world of experience. On that understanding,
we may provisionally distinguish between four domains to
which we may refer when we speak of the *Theory of Statistics*:

(i) *A Calculus of Errors*, as first propounded by Legendre,
Laplace and Gauss, undertakes to prescribe a way of combining
observations to derive a preferred and so-called *best* approxima-
tion to an assumed *true* value of a dimension or constant
embodied in an independently established law of nature. The
algebraic theory of probability intrudes at two levels: (*a*) the
attempt to interpret empirical laws of error distribution
referable to a long sequence of observations in terms consistent
with the properties of models suggested by games of chance;
(*b*) the less debatable proposition that unavoidable observed
net error associated with an isolated observation is itself a
sample of elementary components selected randomwise in
accordance with the assumed properties of such models.

Few current treatises on theoretical statistics have much to say
about the Gaussian Theory of Error; and the reader in search
of an authoritative exposition must needs consult standard
texts on *The Combination of Observations* addressed in the main to
students of astronomy and geodesics. In view of assertions men-
tioned in the opening paragraph of this chapter, it is pertinent
to remark that a calculus for combining observations as propounded by Laplace and by Gauss, and as interpreted by all
their successors, presupposes a putative true value of any
measurement or constant under discussion as a secure foothold

för the concept of error. When expositors of the contemporary reorientation of physical theory equate the assertion that the canonical form of the scientific law is statistical to the assertion that the new physicist repudiates absolute truth, cause and effect as irrelevant assumptions, it is therefore evident that they do not use the term statistical to cover the earliest extensive application of the theory of probability in the experimental sciences. I cannot therefore share with my friend Dr. Bronowski the conviction that the statistical formulation of particular scientific hypotheses subsumed under the calculus of aggregates as defined below has emancipated science from an aspiration so old-fashioned as the pursuit of absolute truth. Still less do I derive from so widely current a delusion any satisfaction from the promise of a new Elizabethan era with invigorating prospects of unforeseen mental adventure.

(ii) *A Calculus of Aggregates* proceeds deductively from certain axioms about the random behaviour of subsensory particles to the derivation of general principles which stand or fall by their adequacy to describe the behaviour of matter in bulk. In this context our criterion of adequacy is the standard of precision we commonly adopt in conformity with operational requirements in the chosen field of enquiry. Clerk Maxwell's Kinetic Theory of Gases is the *fons et origo* of this prescription for the construction of a scientific hypothesis and the parent of what we now call statistical mechanics. Beside it, we may also place the Mendelian Theory of Populations.

So far as I know, the reader in search of an adequate account of statistical hypothesis of this type will not be able to find one in any standard current treatise ostensibly devoted to Statistical Theory as a whole. This omission is defensible in so far as physicists and biologists admittedly accept the credentials of such hypotheses on the same terms as they accept hypotheses which make no contact with the concept of probability, e.g. the thermodynamic theory of the Donnan membrane equilibrium. We assent to them because they *work*. Seemingly, it is unnecessary to say more than that, since all scientific workers agree about what they mean in the laboratory, when they say that a hypothesis works; but such unanimity has no bearing on the plea that statistical theory works, when statistical

we do not mean what we now commonly call *theoretical statistics*
What we do subsume under the latter presupposes that our
analysis invokes the calculus of probabilities. When we speak
of a calculus of probabilities we also presuppose a single formal
system of algebra; but a little reflection upon the history of the
subject suffices to remind us that: (*a*) there has been much
disagreement about the relevance of such a calculus to every-
day life; (*b*) scientific workers invoke it in domains of discourse
which have no very obvious connexion. When I say this I wish
to make it clear that I do not exclude the possibility that we
may be able to clarify a connexion if such exists, but only if we
can reach some agreement about the relevance of the common
calculus to the world of experience. On that understanding,
we may provisionally distinguish between four domains to
which we may refer when we speak of the *Theory of Statistics:*

(i) *A Calculus of Errors*, as first propounded by Legendre,
Laplace and Gauss, undertakes to prescribe a way of combining
observations to derive a preferred and so-called *best* approxima-
tion to an assumed *true* value of a dimension or constant
embodied in an independently established law of nature. The
algebraic theory of probability intrudes at two levels: (*a*) the
attempt to interpret empirical laws of error distribution
referable to a long sequence of observations in terms consistent
with the properties of models suggested by games of chance;
(*b*) the less debatable proposition that unavoidable observed
net error associated with an isolated observation is itself a
sample of elementary components selected randomwise in
accordance with the assumed properties of such models.

Few current treatises on theoretical statistics have much to say
about the Gaussian Theory of Error; and the reader in search
of an authoritative exposition must needs consult standard
texts on *The Combination of Observations* addressed in the main to
students of astronomy and geodesics. In view of assertions men-
tioned in the opening paragraph of this chapter, it is pertinent
to remark that a calculus for combining observations as propounded by Laplace and by Gauss, and as interpreted by all
their successors, presupposes a putative true value of any
measurement or constant under discussion as a secure foothold

and other mathematicians *au fait* with the factual assumptions on which the Gaussian theory relies.

Since an onslaught by Keynes, the name of Quetelet, so explicitly and repeatedly acknowledged by Galton, by Pearson and by Edgworth as the parent of regression and cognate statistical devices, has unobtrusively retreated from the pages of statistical textbooks; but mathematicians of an earlier vintage than Pearson or Edgworth had no illusions about the source of ' 's theoretical claims nor about the relevance of the principle invoked to the end in view. So discreet a disinclinat is v be into the beginnings of much we now encou ..uon, and o. Quetelet s ..... social phenomena are subject to quantitative laws as inexorable as those of Kepler. Procedures we may designate in this way, more especially the analysis of covariance, may invoke significance tests, and therefore intrude into the domain of the calculus of judgments; but the level at which the theory of probability is ostensibly relevant to the original terms of reference of such exploratory techniques is an issue *sui generis*.

The pivotal concept in the algebraic theory of regression, as in the Gaussian theory of error, is that of *point-estimation*; and the two theories are indeed formally identical. In what circumstances this extension of the original terms of reference of the calculus of errors took place is worthy of comment at an early stage. We shall later see how a highly debatable transference of the theory of wagers in games of chance to the uses of the life table for assessment of insurance risks whetted the appetite for novel applications of the algebraic theory of probability in the half-century before the more mature publications of Gauss appeared in print. The announcement of the Gaussian theory itself coincided with the creation of new public instruments for collection of demographic data relevant to actuarial practice both in Britain* and on the Continent. Therewith we witness the emergence of the professional statistician in search of a theory. Such was the setting in which Quetelet obtained a considerable following, despite the derision of Bertrand (p. 172)

* The Office of the Registrar-General of England and Wales came into being in 1837.

theory signifies the contents of contemporary manuals setting forth a regimen of interpretation now deemed to be indispensable to the conduct of research in the biological and social sciences.

(iii) *A Calculus of Exploration*, which I here use to label such themes as regression and factor analysis, is difficult to define without endorsing or disclaiming its credentials. The expression is appropriate in so far as the ostensible aim of procedures subsumed as such is definable in the idiom of Karl Pearson as *concise statement* of unsuspected *regularities of nature*. This aga͙ͅ to pro͙but less so if we interpret it in terms of Pearson's͙ ͙ter in an up-to-date treatise on statistical theory will not necessarily puzzle the reader, if sufficiently acquainted with the unresolved difficulties of reaching unanimity with respect to the credentials of the remaining topics there dealt with; but we shall fail to do justice to the legitimate claims of its contents, unless we first get a backstage view of otherwise concealed assumptions. In what seems to be one of the first manuals setting forth the significance test drill, Caradog Jones (*First Course in Statistics*, 1921) correctly expounds as follows the teaching of Quetelet and its genesis as the source of the tradition successively transmitted through Galton, Pearson and R. A. Fisher to our own contemporaries:

It is almost true to say, however, that until the time of the great Belgian, Quetelet (1796–1874), no substantial theory of statistics existed. The justice of this claim will be recognized when we remark that it was he who really grasped the significance of one of the fundamental principles—sometimes spoken of as the *constancy of great numbers*—upon which the theory is based. A simple illustration will explain the nature of this important idea: imagine 100,000 Englishmen, all of the same age and living under the same normal conditions—ruling out, that is, such abnormalities as are occasioned by wars, famines, pestilence, etc. Let us divide these men at random into ten groups, containing 10,000 each, and note the age of every man when he dies. Quetelet's principle lays down that, although we cannot foretell how long any particular individual will live, the ages at death of the 10,000 added together, whichever group we consider, will be practically the same. Depending upon this fact, insurance companies calculate the premiums they must charge, by a process of averaging mortality results recorded in the past, and so they are able to carry on business without serious risk